Tall Wanda was the biggest, smartest person around. She also had a bull called Watson. Of all bulls, Watson was the strongest. He could squash a tree with one hoof.

One fall day, Wanda saw Big Shot Wally. He was washing his feet in brook water.

"See Squall Mountain?" Wally said. "A lovely lake is beyond. Only one person has seen it."

"Long ways over that mountain," Wanda said.

“I can get to that lake side faster than you can!” Wally bragged. “If you can get there first, I will give you fifty dollars!”

“All right,” Wanda said, “but I always take Watson with me.”

Wally snickered. He felt sure Watson would fall and slow her down.

They said they would start at dawn. Then Wanda washed Watson down and gave him some water. She also got a great big spade and a large wagon.

At dawn, Wally went to Squall Mountain.

"Wanda is late!" Wally said. "I will get there first! Then I am going to name that lake after me. I will call it Wally Lake."

While Wally was climbing, Wanda was digging. She dug huge spadefuls of dirt out of Squall Mountain. She put them in the wagon, and Watson hauled them away. Soon Wanda had dug an immense tunnel.

Then Wanda and Watson walked through the tunnel.

There was the lake, with shining blue water. Watson washed his hoofs while Wanda watched for Wally.

When Wally straggled in, he spotted Wanda's tunnel.

"Not fair!" Wally yelled.

"You said all I had to do was get here first," Wanda said. "You never said how."

"You owe me fifty dollars," said Wanda.

"What about twenty-five?" Wally begged, taking out a small wallet.

"Not enough," grumbled Wanda.

So Wally gave Wanda all he owed. Then Wanda named the lake. What do you think she called it?